Primary Word Problems, Book 2

Primary Word Problems, Book 2

THINKING MATHEMATICALLY

At Home and At School Problem-Solving Ideas for Grades 3-5

Dr. Now

Copyright © 2014 by Dr. Now.

Library of Congress Control Number: 2014920512
ISBN: Hardcover 978-1-5035-1717-2
Softcover 978-1-5035-1719-6
eBook 978-1-5035-1718-9

All rights reserved. No part of this book may be reproduced or transmitted in any form or by any means, electronic or mechanical, including photocopying, recording, or by any information storage and retrieval system, without permission in writing from the copyright owner.

Any people depicted in stock imagery provided by Thinkstock are models, and such images are being used for illustrative purposes only.
Certain stock imagery © Thinkstock.

Print information available on the last page.

Rev. date: 04/23/2015

To order additional copies of this book, contact:
Xlibris
1-888-795-4274
www.Xlibris.com
Orders@Xlibris.com

To Willie Kizer and Hermena Crum

They have always encouraged me to follow my dreams.

To my husband, Johnny Henry Brown

He gives of himself selflessly and helps bring my dream to fruition.

To Dr. Rebecca Dargan

Over the years, she has given tireless effort,
time, and encouragement and
a new meaning to a "guardian angel."

To my sons, Michael and Jason Simmons

They have supported me spiritually, mentally,
physically, and financially
to achieve this dream.

To the Crums, Simmons, and friends,

They have given me the courage to face adversity
and to remember who is really in charge.

Day 1

Your favorite uncle is driving in from Del City, Oklahoma, to your home next week. He's coming from Del City, Oklahoma, to Walterboro, South Carolina. Your uncle lives 1,083.6 miles away from you. Approximately how many hours will it take your uncle to get home if his driving speed is 70 mph? It will take him approximately _____ hours to get home.	Work Space

Day 2

You plan to share a pizza with two buddies. Just before you cut the pizza, a fourth buddy comes to join you. Show and explain what you will do for you and your buddies to have equal slices of the pizza.	Work Space

Day 3

Your favorite uncle gave you $75.00. You are going to try to spend every cent. Your choices are: 　　a pad case, $31.00 　　a bowling ticket, $24.00 　　an aquarium $29.00 　　a video game $44.00 What will you purchase to come as close as possible to using all of your $75.00? I will purchase a/an _____ and a/an _____.	Work Space

Day 4

Fifteen children are in Dr. Now's class. Nine of the children are girls. How many boys are in Dr. Now's class? There are ____ boys in Dr. Now's class.	Work Space

Day 5

If Saturday school begins at 9:00 a.m. and ends three hours later, at what time will Saturday school end? Saturday School will end at _____.	Work Space

Day 6

There were 14 children at a party. There were 36 cupcakes. If each child ate two cupcakes, how many cupcakes were left? There were ____ cupcakes left.	Work Space

Day 7

I can pick 50 tomatoes in 2 hours. How many tomatoes can I pick in 8 hours? I can pick _____ tomatoes in 8 hours.	Work Space

Day 8

There are 289 sheep in the field. Each sheep has four legs. How many legs do the 289 sheep have all together? There are 289 sheep in the field with _____ legs all together.	Work Space

Day 9

There are 165 calories in 15 potato chips. How many calories are there in each potato chip? Each potato chip has ___ calories.	Work Space

Day 10

A daycare center ordered 244 new blankets. The blankets will be delivered in 4 equal shipments. How many blankets will be in each shipment? There will be ___ blankets in each shipment.	Work Space

Day 11

The marching band has 200 students. How many rows of students can be formed with 8 students in each row? There are ___ rows of students in the marching band.	Work Space

Day 12

The soda factory filled 7,432 bottles in 2 hours. How many bottles were filled every half hour (30 minutes)? There were _____ bottles filled every half hour.	Work Space

Day 13

Mr. Simmons has 588 cement blocks. He is building a storage shed. How many rows of 12 cement blocks can he lay for the storage shed? He can lay _____ rows of cement blocks for the storage shed.	Work Space

Day 14

Ethan received a new video game for his birthday. He scored 4 points on the first level, 16 points on the second level, and 64 points on the third level. If this pattern continues, how many points will Ethan score on the fifth level? Ethan will score _____ on the fifth level.	Work Space

Day 15

Malachi has a marble collection and likes to display his collection in trays. He puts 3 marbles in the first tray, 5 marbles in the second tray, 8 marbles in the third tray, and 12 marbles in the fourth tray. If this pattern continues, how many marbles will Malachi display in the fifth tray? Malachi will display _____ marbles in the fifth tray.	Work Space

Day 16

While putting potatoes in piles for the potato bank, Uncle George put 1 potato in the first pile, 3 potatoes in the second pile, 5 potatoes in the third pile, and 7 potatoes in the fourth pile. If he continues this pattern, how many potatoes will he put in the fifth pile? He will put _____ potatoes in the fifth pile.	Work Space

Day 17

While sorting some hair barrettes, Elizabeth drops 21 barrettes in the first jar, 28 barrettes in the second jar, 35 barrettes in the third jar, 42 barrettes in the fourth jar, and 49 barrettes in the fifth jar. If she continues this pattern, how many barrettes will Elizabeth drop in the sixth jar? Elizabeth will drop _____ barrettes in the sixth jar.	Work Space

Day 18

Approximately how many centimeters equal 2 inches? Around _____ centimeters equal 2 inches.	Work Space

Day 19

Approximately how many quarts are equal to 5 liters? Around _____ quarts are equal to 5 liters.	Work Space

Day 20

Myron has a piece of green yarn and a piece of yellow yarn. The piece of green yarn is 5 inches shorter than the piece of yellow yarn. Myron also has 7 inches of scotch tape. The piece of yellow yarn is 29 inches long. How many inches of yarn does Myron have in all? Myron has _____ inches of yarn in all.	Work Space

Day 21

How many years are in a fourth of a century? There are _____ years in a fourth of a century.	Work Space 1 century equals 100 years.

Day 22

How many minutes are in 1/10 of an hour? There are _____ minutes in 1/10 of an hour.	Work Space

Day 23

How many years are in ½ of a decade? There are _____ years in ½ of a decade.	Work Space

Day 24

How any hours is a fourth of a day? There are _____ hours in a fourth of a day.	Work Space

Day 25

Gabrielle has 10 pink ribbons. She has 8 fewer green ribbons than pink ribbons. She has 16 more purple ribbons than green ribbons. How many ribbons does Gabrielle have in all? Gabrielle has _____ ribbons in all.	Work Space

Day 26

Ethan saved $54.00. He earned $35.00 for cutting the lawn. Then he spent $10.00 on a pair of goggles, $15.00 on an iPad cover, $14.75 for a sports cap, and $25.00 for a blue scarf. How much money does Ethan have left? Ethan has _____ left.	Work Space

Day 27

In Michael's desk drawer, there are 14 yellow markers. There are 4 more black markers than yellow markers, and there are 11 more green markers than black markers. How many markers are in Michael's desk drawer all together? There are _____ markers in Michael's desk all together.	Work Space

Day 28

Jason planted four types of fruit trees in his vineyard. He planted 37 of each type of fruit. How many fruit trees did Jason plant in his vineyard? Jason planted _____ fruit trees in his vineyard.	Work Space

Day 29

The Simmons' Explorer has 816 campers. If each cabin can hold 8 campers, how many cabins will they need? They will need _____ cabins.	Work Space

Day 30

Elizabeth was collecting sparkling marbles. She got 23 marbles in a fish bowl, and Emmanuel gave her 8 marbles. Her dad gave her 14 marbles. Elizabeth gave 17 marbles to her cousin Alex. How many marbles did Elizabeth have left? Elizabeth had _____ marbles left.	Work Space

Day 31

George wants to triple the size of his work area. He has an area of 123 feet. How much work area does George want? *Do you understand the question?* *Make a plan to solve the problem.* *Carry out your plan to solve the problem.* *Check your answer.*	Work Space

Day 32

Rosie has a rectangular flower garden that is 4 ft long and 6 ft wide. What length of fencing does she need to fence in her flower garden? Rosie needs _____ of fence for fencing in her flower garden. *Do you understand the question?* *Make a plan to solve the problem.* *Carry out your plan to solve the problem.* *Check your answer.*	Work Space

Day 33

Dr. Now bought a one-year fitness contract for $ 420.00. If she makes monthly payment for a year how much is each payment:

Dr. Now's monthly payment will be $_____.

Work Space

Day 34

Your mother wants new carpet in her bedroom. Her bedroom is 12m by 12m rectangle. How much carpet does she needs to buy to cover her bedroom floor?

She needs to buy _____ to cover the entire bedroom floor.

Work Space

Day 35

A daycare center teacher ordered 69 boxes of popsicle sticks. There were 85 popsicle sticks in each box. How many popsicle sticks did the teacher order in all? The teacher ordered _____ popsicle sticks in all.	Work Space

Day 36

The cooks at the shelter need 654 potatoes. There are 6 potatoes in each sack. How many sacks of potatoes do they need? They need _____ sacks of potatoes.	Work Space

Day 37

The school principal has $6,918 to spend on replacing tiles for the classroom. If the cost of each tile is $3.00, how many tiles can the principal buy? The principal can buy _____ tiles.	Work Space

Day 38

A district of teachers needs to go to a workshop in a nearby city. There are 780 teachers in the district. If each passenger van carries 15 teachers, how many passenger vans does the district need? The district needs _____ passenger vans.	Work Space

Day 39

A cheerleader squad has $829.00 to spend on buttons to give their fans. The cost of each button is $0.85. Approximately how many buttons can the squad purchase? The squad can purchase approximately _____ buttons for their fans.	Work Space

Day 40

At the Carolina Fair, a group of 554 seniors wants to ride the giant Tagmahall. If each booth on the Tagmahall can hold 9 seniors in each car, approximately how many cars will the seniors need? The seniors will need approximately ___ cars.	Work Space

Day 41

Your dad spent $287.00 on a diamond necklace and $878.00 on a set of diamond earrings. How much money did he spend in all for your mother's birthday jewelry?

He spent _____ for the birthday jewelry.

Work Space

Day 42

Before lunch, Jason has a science class for 1 hour and 40 minutes. Lunch lasts 45 minutes and ends at 1:00 p.m. What time does his science class start?

His science class starts at _____.

Work Space

Day 43

You need to drink 7 ¼ cups of water each day and ¾ cups of milk each day. Approximately how much liquid do you have to drink daily? I have to drink around _____ of liquid daily.	Work Space

Day 44

The class has 24 students, and ¾ of them are boys. How many are boys in the class? There are _____ boys in the class.	Work Space

Day 45

The cook made 48 cupcakes. One-eighth of the cupcakes have green icing. How many cupcakes are green? There are _____ green cupcakes.	Work Space

Day 46

Three-sixths of the 18 children on the playground are playing kickball. How many children are playing kickball? There are _____ children playing kickball.	Work Space

Day 47

Thirty-six students are eating lunch in the cafeteria. Three-fourths of the students are eating cafeteria lunch. How many students are eating cafeteria lunch? There are _____ students eating the cafeteria lunch.	Work Space

Day 48

There are 50 people waiting for the Greyhound bus. One-fifth of them have two or more pieces of luggage. How many people have two or more pieces of luggage? There are _____ people with two or more pieces of luggage.	Work Space

Day 49

One-fifth of the 895 students at school wear glasses. How many students wear glasses?

There are _____ students who wear glasses.

Work Space

Day 50

There are 750 houses in our community. Two-fifths of the houses are made of bricks. How many houses are made of bricks?

There are _____ houses that are made of bricks.

Work Space

Day 51

Renting a bike at the amusement park costs $12.00 per hour. A family of four rented 4 bikes. How much will they pay to rent the bikes for 6 hours? They will pay _____ to rent the bikes for 6 hours.	Work Space

Day 52

Jason earned $25.00 for grooming the dog. He earned $35.00 for cutting the lawn. His brother Michael earned $65.00 for replacing a ceiling fan. How much money did Jason earn in all? He earned _____ in all.	Work Space

Day 53

Janee walks for 30 minutes each day. She does aerobics for 25 minutes each day. How many minutes does Janee walk and do aerobics in 7 days? Janee walks and does aerobics for _____ minutes in 7 days.	Work Space

Day 54

Tickets to the amusement park cost $12.00 for children and $15.00 for adults. How much will it cost for 6 adults and 9 children to enter the amusement park? *Talk to a partner to discuss the correct order of the operations you need to use to solve the problem: multiply and then subtract; subtract and then multiply; multiply and then add; or add and then multiply.*	Work Space

Day 55

It takes Michael 53 minutes to walk from his house to the store. If Michael needs to be at the store at 8:30 a.m. for work, what time does he need to leave his house? Michael needs to leave his house by _____ a.m. to get to work on time.	Work Space

Day 56

Mrs. Simmons bought a purse and a pair of shoes at the department store. She paid $59.00 for the shoes and $29.99 for the purse. The clerk gave her a change of $1.01. How much money did she give the clerk? She gave the clerk _____.	Work Space

Day 57

Reduce the fraction 4/12 to its lowest term.	Work Space

Day 58

Reduce the fraction 6/8 to its lowest term.	Work Space

Day 59

What is the area of the rectangle?	Work Space
10cm 6cm	

Day 60

What is the perimeter of the square?	Work Space
8 m	

Day 61

What is the perimeter of the rectangle? 6cm ☐ 7cm	Work Space

Day 62

What is the area of the figure below? ☐ 8 in. 3 in.	Work Space

Day 63

Draw a right angle.	Work Space

Day 64

Draw an obtuse triangle.	Work Space

Day 65

A ticket to the talent show costs $3.00. How much will it cost if 5 people go? 7 people? 9 people?

Tickets	1	2	3	4	5	6
Cost	$3	$6	$9	$12	$15	$18

Work Space

Day 66

Reduce 8/12 to the lowest term.

Work Space

Day 67

What two fractions are equivalent to 4/5?	Work Space

Day 68

Divide this pizza into 8 equal parts. ○ Shade 5/8 of the pizza.	Work Space

Day 69

The cat's birthday cake has ¼ cup of milk and ¾ cup of water. How much liquid is needed in the cake? _____ of liquid is needed in the cake.	Work Space

Day 70

James spent 3/6 of an hour studying for his social studies exam and 5/6 of an hour studying for his French test. How much longer did James spend studying for French than for social studies? He spent _____ minutes more studying French than studying Social Studies	Work Space

Day 71

Thea bought ¾ lbs. of green bananas and 2/4 lbs. of yellow (ripe) bananas. How many pounds of bananas did Thea buy in all? Thea bought _____ lbs. of bananas in all.	Work Space

Day 72

Helen baked 252 cookies. She wants to give all six of her grandchildren the same number of cookies. How many cookies should each grandchild get? Helen should give each grandchild _____ cookies.	Work Space

Day 73

Jason received $85.00 from his part-time job. He went to a football game, a movie, and the game room. He had $12.50 left. How much did he spend on entertainment?

Jason spent $ _____ on entertainment.

Work Space

Day 74

Last week, Ethan made $79.00 doing chores. He decided to wash cars at a pit stop. He now has $130.00. How much money did he make washing cars?

Jason made $ _____ washing cars.

Work Space

Day 75

Frederick has 85 watermelons left at his vegetable stand. He sent his workers to get more watermelons to stock for the afternoon. Now 105 watermelons are at his stand. How many did the workers bring from the field? Frederick's workers brought _____ watermelons from the field.	Work Space

Day 76

Draw a clock with the time 1:40.
 	
   a. What time will it be in 2 hours and 30 minutes? _____
   b. What time will it be in 3 hours and 15 minutes? _____
	Work Space

Day 78

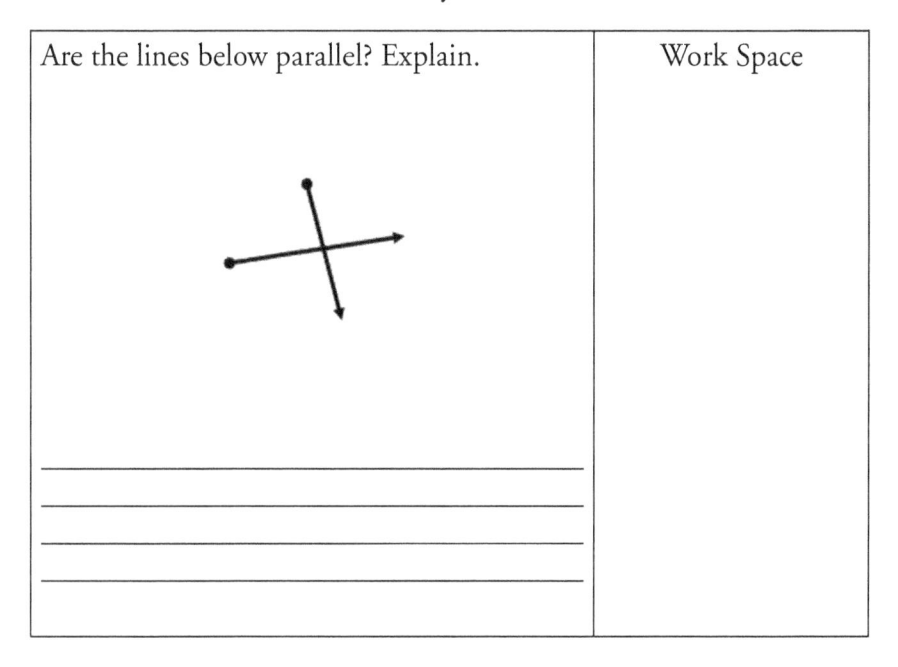

Are the lines below parallel? Explain.	Work Space
_____ _____ _____ _____	

Day 79

Are the lines below parallel? Explain.	Work Space
_____ _____ _____ _____	

Day 80

Are the lines below intersecting? Explain.

_____
_____
_____
_____

Work Space

Day 81

How many vertices does this shape have? Explain.

Work Space

Day 82

How many sides does this shape have?	Work Space
	Sides are the straight lines that make up a shape.
This shape has _____ sides.	

Day 83

What is the unknown number? $7 + r = 11$ $r =$ _____	Work Space

Day 84

Johnny has $24.00 to spend on six ink pens. After buying the pens, he had $12.00 left. How much did each ink pen cost? Each ink pen cost $ _____ .	Work Space

Day 85

There were 324 students going to the zoo. The students filled six buses. How many students were on each bus? There were _____ students on each bus.	Work Space

Day 86

After having lunch at a restaurant, Dee, Rosemary, and Robin decided to evenly split the cost for lunch. If each person paid $19.00, what was the total bill for the lunch? The total bill for the lunch was $_____.	Work Space

Day 87

Converting: How many minutes and hours are in 205 minutes? *Remember: 60 minutes = 1 hour* There are _____ hours and _____ minutes in 205 minutes.	Work Space

Day 88

Converting: How many minutes and hours are in 159 minutes? *Remember: 60 minutes = 1 hour* There are _____ hours and _____ minutes in 159 minutes.	Work Space

Day 89

Converting: How many minutes are in 5 hours? *Remember: 60 minutes = 1 hour* There are _____ minutes in 5 hours.	Work Space

Day 90

Solve the equation: 3 = 6t - 12 - 3 t = _____	Work Space

Day 91

Solve the equation: 36 = -p - 5p p = _____	Work Space

Day 92

Solve the equation: q + 8 - 2 = 9 q = _____	Work Space

Day 93

	Work Space
Make a bar graph with 7 classmates' favorite vegetables. **Favorite Vegetables** squash / green beans / cabbage / carrots / cucumber What is the most favorite vegetable? _____ What is the least favorite vegetable? _____	

Day 94

	Work Space
Create a graph of 10 classmates' favorite subjects. **Favorite Subject** Reading / Math / Science / Social Studies / Writing What is the most favorite subject? _____ What is the least favorite subject? _____	

Day 95

Make a bar graph using a cup of colorful marshmallows and create two questions about your graph. Graphing _____ _____ _____	Work Space

Day 96

Solve the equation: 4 X 27 ÷ 3 = _____	Work Space

Day 97

Solve the equation:	Work Space
(20 X 3) ÷ 3 = _____	

Day 98

Solve the equation:	Work Space
3.07 + 5. 4 - 2.1 = _____	

Day 99

Using decimals:

The vet is 7.8 kilometers west of the gas station and 25.6 kilometers east of the beauty salon. How far apart are the gas station and the beauty salon?

The gas station and the beauty salon are _____ kilometers apart.

Work Space

Day 100

Using decimals:

The mechanic shop is 71.6 miles west of the donut shop and 54.5 miles east of the barber shop. How far is the donut shop from the barber shop?

The donut shop is _____ miles from the _____ shop.

Work Space

Day 101

Rachael is reading a mystery novel that has 581 pages. She wants to finish reading the novel in seven days. How many pages does she need to read each day? She needs to read _____ pages each day.	Work Space

Day 102

The Parent and Teacher Organization (PTO) bought 14 boxes of lead pencils. Each box has 18 lead pencils. The PTO paid $6.00 per box. If the PTO sells each lead pencil for $0.50, how much profit will the PTO make? The PTO will make a _____ profit.	Work Space

Day 103

Jason and his friends play baseball on Saturday. The game begins at 2:30 p.m. It will take Jason and his friends 40 minutes to get to the field. Approximately what time do they need to leave in order to get to the baseball field 30 minutes before the game begins? They need to leave at _____.	Work Space

Day 104

Michael's football practice began at 11:00 a.m. The team practiced offense for 65 minutes, and 1 hour and 45 minutes on defense techniques. What time did Michael's practice end? Michael's practice ended at _____.	Work Space

Day 105

Elizabeth went to the park at 2:35 p.m. She played in the sandbox for 30 minutes, the slides for 10 minutes, and the swings for 15 minutes. What time did Elizabeth leave the park? Elizabeth left the park at _____.	Work Space

Day 106

On Tuesday, Emmanuel had a student council meeting after class for 45 minutes, followed by keyboarding music lessons for 1 hour and 10 minutes. If Emmanuel's classes end at 3:15 p.m., what time was he finished with his music lesson? He finished his music lesson at _____.	Work Space

Day 107

John bought 43 cases of carnation flowers. There were 35 carnations in each case. How many carnations did John buy? John bought _____ carnations.	Work Space

Day 108

The school district has 65 buses. There are 39 seats on each bus. Approximately how many children can the buses transport to school? The buses can transport _____ children to school.	Work Space

Day 109

At Elizabeth's birthday party, the girls ate 2½ pizzas and the boys ate 4¾ pizzas. How many pizzas did the boys and girls eat in all?

The boys and the girls ate _____ pizzas in all.

Work Space

Day 110

Mr. Johnny bought 5 gallons of paint. He used only 1 ¾ gallon. How much paint does Mr. Johnny have left?

Mr. Johnny has _____ gallons of paint left.

Work Space

Day 111

My German Shepherd is 6½ years old, and my Cocker Spaniel is 2½ years younger than the German Shepherd. How old is my Cocker Spaniel? The Cocker Spaniel is _____ years old.	Work Space

Day 112

Shade the parts to illustrate the decimal number 0.4.	Work Space

Day 113

Shade the parts to illustrate the decimal number 0.74.	Work Space

Day 114

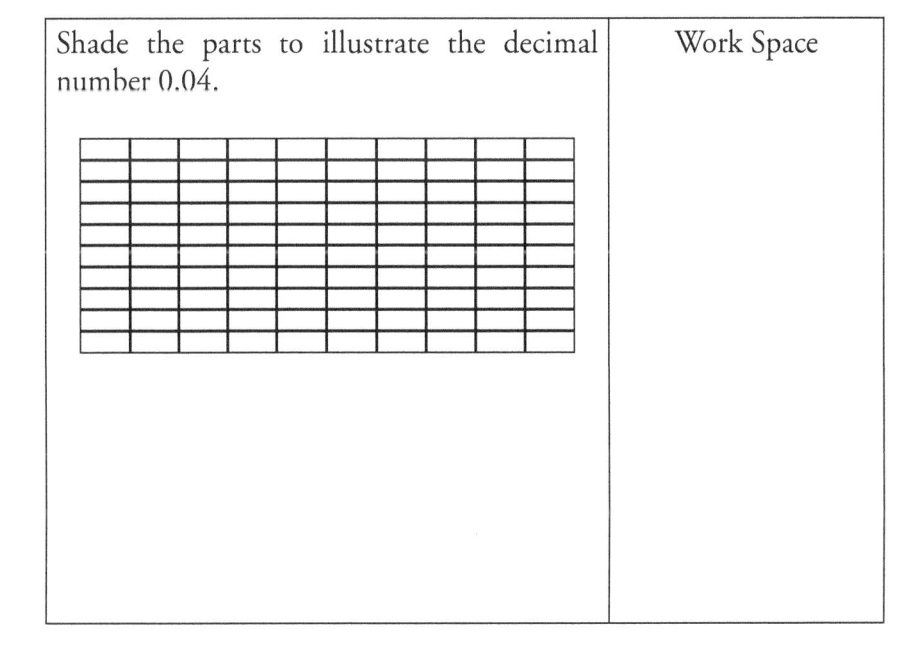

Shade the parts to illustrate the decimal number 0.04.	Work Space

Day 115

Decimals: Janee bought 5.1 meters of cloth to cover a scrapbook. She used 4.3 meters of the cloth. How much cloth did Janee have left? Janee had _____ meters of cloth left.	Work Space

Day 116

Jason had 0.9 grams of salt. He used only 0.5 grams of salt to make chicken fritters. How much salt did Jason have left? Jason had _____ grams of salt left.	Work Space

Day 117

Probability:	Work Space
Use words like **certain, unlikely, impossible,** and **probably** to solve problems. Illustrate 3 blue marbles and 2 yellow marbles. How likely is it that you will pick a blue marble out of a bag? Explain.	

Day 118

Probability:	Work Space
Use words like **certain, unlikely, impossible,** and **probably** to solve problems. Illustrate 5 yellow rocks and 4 brown rocks. How likely is it that you will pick a yellow rock?	

Day 119

Place Value:	Work Space
Looking at the number 3,485, which digit is in the tens place? _____ is in the tens place.	

Day 120

Even Numbers: Write the even numbers from 44 to ___. 44, ___, ___, ___, ___, ___, ___, ___, ___, ___, ___, ___, ___, ___, ___, ___, ___, ___, ___, ___	Work Space

Day 121

Odd Numbers: Write the odd numbers from 1 to ___. 1, ___, ___, ___, ___, ___, ___, ___, ___, ___, ___, ___, ___, ___, ___, ___, ___, ___, ___, ___, ___, ___, ___, ___, ___, ___, ___	Work Space

Day 122

Counting by 2s. 50, ___, ___, ___, ___, ___, ___, ___, ___, ___, ___, ___, ___, ___, ___, ___, ___, ___, ___, ___, ___, ___, ___, ___, ___, ___, ___	Work Space

Day 123

Counting by 4s from 44.

44, ___, ___, ___, ___, ___, ___,

___, ___, ___, ___, ___, ___, ___,

___, ___, ___, ___, ___, ___, ___,

___, ___, ___, ___, ___, ___, ___

Work Space

Day 124

Counting by 10s from 43.

43, ___, ___, ___, ___, ___, ___,

___, ___, ___, ___, ___, ___, ___,

___, ___, ___, ___, ___, ___, ___,

___, ___, ___, ___, ___, ___, ___

Work Space

Day 125

Counting by 10s from 112.	Work Space
112, ___, ___, ___, ___, ___, ___, ___, ___, ___, ___, ___, ___, ___, ___, ___, ___, ___, ___, ___, ___, ___, ___, ___, ___, ___, ___, ___	

Day 126

Counting by 5s from 102.	Work Space
102, ___, ___, ___, ___, ___, ___, ___, ___, ___, ___, ___, ___, ___, ___, ___, ___, ___, ___, ___, ___, ___, ___, ___, ___, ___, ___, ___	

Day 127

Number Words: Write the number word for 587.	Work Space
_____ _____ _____ _____	

Day 128

Number Words: Write the number word for 1,802.	Work Space
_____ _____ _____ _____	

Day 129

Ethan bought a 5-lb. bag of apples for $4.38, and he gave the cashier $5.00. How much change will he get? He will get a change of _____.	Work Space

Day 130

You went to the store to get a gallon of vanilla ice cream. The ice cream costs $5.89. Your mom gave you 3 one-dollar bills, 8 quarters, 5 dimes, and 6 pennies. How much money did your mom give you? _____ Do you have enough money to get the ice cream? Yes _____ or No _____ How much money do you have left or how much more do you need? _____	Work Space

Day 131

We are planting 48 sweet potato plants and would like for them to be in 6 rows. How many sweet potato plants need to be in each row?	Work Space

Day 132

At your lemonade stand, you sold lemonade for $0.15 a cup. You made $18.00. How many cups of lemonade did you sell? I sold _____ cups of lemonade.	Work Space

Day 133

The PTO sold T-shirts for $10.00 each to 856 students at school. Everyone purchased a T-shirt for $10.00. How much money did the PTO make in selling the T-shirts? _____ The PTO paid $6.25 for each T-shirt. How much profit did the PTO make? _____	Work Space

Day 134

Mrs. Frederick made a batch of cookies. There were 24 cookies in each batch. She ate ¼ of the batch. Then Mrs. Frederick baked three more batches. How many cookies does Mrs. Frederick have now? Mrs. Frederick has _____ cookies now.	Work Space

Day 135

Emmanuel's grades in science were 89, 93, 87, 77, 100, and 85. What is his average grade in science? Emmanuel's average grade in science is _____.	Work Space

Day 136

Angie wants to give the server a 15% tip. Her lunch receipt is $18.89. How much will Angie give the server? Angie will give the server _____.	Work Space

Day 137

A rectangular court measures 12 ft. by 5 ft. What is the area of the court? The area of the court is _____.	Work Space

Day 138

A square room measures 8 ft. on the right side. What is the perimeter? The perimeter of the room is _____.	Work Space

Day 139

George is putting a new carpet in his bedroom. His bedroom is 21 ft. by 17 ft. What is the area of his bedroom? The area of his bedroom is _____.	Work Space

Day 140

Dr. Dargan has a rose garden that is 8 m by 12 m. One bag of horse manure can cover 12 m. How many bags of horse manure will Dr. Dargan use? Dr. Dargan will use _____ bags of horse manure.	Work Space

Day 141

Elapsed time:	Work Space
If it is 8:15 a.m., what time will it be in 7 hours and 15 minutes? It will be _____ in 7 hours and 15 minutes.	

Day 142

Elapsed time:	Work Space
If it is 1:45 p.m., what time will it be in 7 hours and 30 minutes? It will be _____ in 7 hours and 30 minutes.	

Day 143

Mark started baking some chocolate chip cookies at 4:35 p.m. It took 35 minutes to prepare the ingredients and 2 hours and 10 minutes to bake the cookies. When did Mark finish baking? Mark finished baking at _____.	Work Space

Day 144

Ethan and Elizabeth were watching a video at 5:35 p.m. The video lasted for 1 hour and 45 minutes. Then Ethan and Elizabeth played basketball for 35 minutes. What time did Elizabeth and Ethan finish playing basketball? Elizabeth and Ethan finished playing basketball at _____.	Work Space

Day 145

What shape has three equal sides? Illustrate and name the shape.	Work Space

Day 146

What shape has four equal sides and four corners? Illustrate and name the shape.	Work Space

Day 147

What shape has six sides and six angles? Illustrate and name the shape.	Work Space

Day 148

What shape has eight sides? Illustrate and name the shape.	Work Space

Day 149

Convert the number of miles it takes to equal 15,840 feet. _____ miles equals 15,840 feet.	Work Space

Day 150

How much time elapsed between the first and second time? *First* *Second* *Elapsed Time* 1:45 2:30 _____ 4:15 5:46 _____ 9:35 9:45 _____	Work Space

Day 151

Thea has 895 erasers in a box. She sold 186 erasers. How many does she have left? She has _____ erasers left.	Work Space

Day 152

Lillian has 834 dimes in her piggy bank. She spent 479 dimes. How many dimes does she have now? She has _____ dimes now.	Work Space

Day 153

Lula sold 819 cooking magazines on Monday, and sold an additional 749 magazines on Tuesday. How many more magazines did she sell on Monday than she sold on Tuesday?

She sold _____ more magazines on Monday than she did on Tuesday.

Work Space

Day 154

A toy factory made 589,877 trucks and 539,227 dolls. How many more trucks than dolls did the toy factory make?

The factory made _____ more trucks than dolls.

Work Space

Day 155

During the midnight shift, the workers made 5,277 small bobby pins and 3,456 large bobby pins. How many bobby pins did the workers make during the midnight shift? The workers made _____ bobby pins during the midnight shift.	Work Space

Day 156

Last year, a candy factory made 6,899 chocolate Easter eggs. This year, the candy factory made 10,100 chocolate Easter eggs. How many more chocolate eggs did they make this year than last year? They made _____ more chocolate eggs this year than last year.	Work Space

Day 157

Simmons and Brown Enterprise sold 6,990 paperback books. The company also sold 3,011 eBooks. How many books did Simmons and Brown Enterprise sell in all? Simmons and Brown enterprise sold _____ books in all.	Work Space

Day 158

The Crum's Foundation cut down 4,660 trees on the north side of the estate. They cut down 5,896 more from the south side of the estate. How many more trees were cut down from the south side than the north side? There were _____ more trees cut down from the south side than the north side.	Work Space

Day 159

Wilhemenia wants to build a fence around her backyard. Her backyard is 15 meters long and 15 meters wide. It costs $7.00 per meter to install the fence. How much will it cost to put a fence around Wilhemenia's backyard? It will cost _____ to put a fence around Wilhemenia's backyard.	Work Space

Day 160

The CEO wants to install a new floor in the office. The office measures 9 meters wide and 12 meters long. The flooring costs $8.00 per square meter. How much will the CEO pay for the new floor in the office? The CEO will pay _____ for the new floor in the office.	Work Space

Day 161

Patterns: 1, 3, 9, ___, 81, 243 1, 2, 4, ___, 16, 32 1, 3, 9, ___, 81	Work Space

Day 162

The Ridgeville Community Resource Center used a grant to buy 13,456 books for preschool children. The center distributed 8,045 of the books. How many books does the center have left? The center has _____ books left.	Work Space

Day 163

The staff at the Native American Center treated 879 clients this month. Last month, the staff treated 1,245 clients. How many more clients did the staff treat last month than this month? The staff treated _____ more clients last month than this month.	Work Space

Day 164

What is the median of each of the two sets of numbers? Set 1: 3 5 5 7 8 9 9 Set 2: 1 2 2 4 5 6 8 The median of set 1 is _____. The median of set 2 is _____.	Work Space

Day 165

What is the mean of each of the two sets of numbers? Set 1: 1 6 7 3 8 9 3 3 Set 2: 3 5 4 9 7 9 6 5 The mean of set 1 is _____. The mean of set 2 is _____.	Work Space

Day 166

What is the median of each of the two sets of numbers? Set 1: 8 8 7 6 5 4 1 Set 2: 9 6 4 7 8 3 2 The median of set 1 is _____. The median of set 2 is _____.	Work Space

Day 167

What is the mode of each of the two sets of numbers?

Set 1:
9 5 6 5 3 5 6 4

Set 2:
0 0 7 9 2 1 0 3

The mode of set 1 is _____.
The mode of set 2 is _____.

Work Space

Day 168

What is the mode of each of the two sets of numbers?

Set 1:
8 9 6 8 5 8

Set 2:
7 7 7 9 9 3 3

The mode of set 1 is _____.
The mode of set 2 is _____.

Work Space

Day 169

What is the range of each of the two sets of numbers? Set 1: 5 9 4 8 9 4 3 Set 2: 3 7 5 10 4 2 7 9 The range of set 1 is _____. The range of set 2 is _____.	Work Space

Day 170

Angie bought a bag of speckled beans. The clerk asked for $1.79 for the bag of speckled beans. What coins can she give the clerk to pay the cost of the speckled beans? She can give the clerk _____ _____.	Work Space

Day 171

Wanda bought a container of eggs for $2.97. She gave the cashier $5.00. How much change will she get back? She will get a change of _____.	Work Space

Day 172

Jason bought a pack of fat-free cheese for $3.99. He has 2 dollars, 4 quarters, and 5 dimes. How much change will Jason get back or how much more money will he need to buy the fat-free cheese?	Work Space

Day 173

Write a word problem with your partner and design a strategy to get an answer.	Work Space

Day 174

Write a word problem with your partner and design a strategy to get an answer.	Work Space

Day 175

Write a word problem with your partner and design a strategy to get an answer.	Work Space

Day 176

Write a word problem with your partner and design a strategy to get an answer.	Work Space

Day 177

Write a word problem with your partner and design a strategy to get an answer.	Work Space

Day 178

Write a word problem with your partner and design a strategy to get an answer.	Work Space

Day 179

Write a word problem with your partner and design a strategy to get an answer.	Work Space

Day 180

Write a word problem with your partner and design a strategy to get an answer.	Work Space

ANSWER KEY

Day 1

Your favorite uncle is driving in from Del City, Oklahoma, to your home next week. He's coming from Del City, Oklahoma, to Walterboro, South Carolina. Your uncle lives 1,083.6 miles away from you. Approximately how many hours will it take your uncle to get home if his driving speed is 70 mph? It will take him approximately 16 hours to get home.	Work Space 70 mph = 15hours and 48 minutes

Day 2

You plan to share a pizza with two buddies. Just before you cut the pizza, a fourth buddy comes to join you. Show and explain what you will do for all of you to have equal slices of the pizza.	Work Space ◯

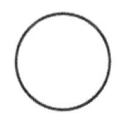

Day 3

Your favorite uncle gave you $75.00. You are going to try to spend every cent. Your choices are: a pad case, $31.00, a bowling ticket, $24.00, an aquarium $29.00, a video game $44.00. What will you buy to come close to using all of your $75.00?	Work Space The answers may vary.

Day 4

Fifteen children are in Dr. Now's class. Nine of the children are girls. How many boys are in Dr. Now's class? There are 6 boys in Dr. Now's class.	Work Space 15 – 9 = 6

Day 5

If Saturday school begins at 9:00 a.m. and ends three hours later, at what time will Saturday school end? Saturday School will end at 12:00 p.m.	Work Space 9 + 3 = 12

Day 6

There were 14 children at a party. There were 36 cupcakes. If each child ate two cupcakes, how many cupcakes were left? There were 8 cupcakes left.	Work Space 14 x 2 = 28 36 – 28 = 8

Day 7

I can pick 50 tomatoes in 2 hours. How many tomatoes can I pick in 8 hours? I can pick 200 tomatoes in 8 hours.	Work Space 50 + 50 + 50 + 50 = 200 or 25 X 8 = 200

Day 8

There are 289 sheep in the field. Each sheep has four legs. How many legs do the 289 sheep have all together? There are 289 sheep in the field with 1,156 legs all together.	Work Space 289 X 4 = 1,156 legs

Day 9

There are 165 calories in 15 potato chips. How many calories are there in each potato chip? Each potato chip has 11 calories.	Work Space 165 ÷ 15 = 11

Day 10

A daycare center teacher ordered 244 new blankets. The blankets will be delivered in 4 equal shipments. How many blankets will be in each shipment? There will be 61 blankets in each shipment.	Work Space 244 ÷ 4 = 61

Day 11

The marching band has 200 students. How many rows of students can be formed with 8 students in each row? There are 25 rows of students in the marching band.	Work Space 200 ÷ 8 = 25

Day 12

The soda factory filled 7,432 bottles in 2 hours. How many bottles were filled every half hour (30 minutes)? There were 1,858 bottles filled every half hour.	Work Space 7,432 ÷ 4 = 1,858

Day 13

Mr. Simmons has 588 cement blocks. He is building a storage shed. How many rows of 12 cement blocks can he lay for the storage shed? He can lay 49 rows of cement blocks for the storage shed.	Work Space 588 ÷ 12 = 49

Day 14

Ethan received a new video game for his birthday. He scored 4 points on the first level, 16 points on the second level, and 64 points on the third level. If this pattern continues, how many points will Ethan score on the fifth level? Ethan will score 1024 points on the fifth level.	Work Space 256 X 4 = 1024

Day 15

Malachi has a marble collection and likes to display his collection in trays. He puts 3 marbles in the first tray, 5 marbles in the second tray, 8 marbles in the third tray, and 12 marbles in the fourth tray. If this pattern continues, how many marbles will Malachi display in the fifth tray? Malachi will display 17 marbles in the fifth tray.	Work Space 12 + 5 = 17

Day 16

While putting potatoes in piles for the potato bank, Uncle George put 1 potato in the first pile, 3 potatoes in the second pile, 5 potatoes in the third pile, and 7 potatoes in the fourth pile. If he continues this pattern, how many potatoes will he put in the fifth pile? He will put 9 potatoes in the fifth pile.	Work Space pattern odd numbers 1,3,5,7,9

Day 17

While sorting some hair barrettes, Elizabeth drops 21 barrettes in the first jar, 28 barrettes in the second jar, 35 barrettes in the third jar, 42 barrettes in the fourth jar, and 49 barrettes in the fifth jar. If she continues this pattern, how many barrettes will Elizabeth drop in the sixth jar? Elizabeth will drop 56 barrettes in the sixth jar.	Work Space 49 + 7 = 56

Day 18

Approximately how many centimeters equal 2 inches? Around 5 centimeters equal 2 inches.	Work Space 5 centimeters is about equal to 2 inches

Day 19

Approximately how many quarts are equal to 5 liters? Around 5 quarts are equal to 5 liters.	Work Space 1 liter is about 1 quart; therefore, 5 liters is approximately 5 quarts.

Day 20

Myron has a piece of green yarn and a piece of yellow yarn. The piece of green yarn is 5 inches shorter than the piece of yellow yarn. Myron also has 7 inches of scotch tape. The piece of yellow yarn is 29 inches long. How many inches of yarn does Myron have in all? Myron has 53 inches of yarn in all.	Work Space (29 - 5) + 29 = 53 inches

Day 21

How many years are in a fourth of a century? There are 25 years in a fourth of a century.	Work Space 1 century equals 100 years. 100 ÷ 4 = 25

Day 22

How many minutes are in 1/10 of an hour?	Work Space
There are 6 minutes in 1/10 of an hour.	60 ÷ 10 = 6

Day 23

How many years are in ½ of a decade?	Work Space
There are 5 years in ½ of a decade.	10 ÷ 2 = 5

Day 24

How any hours is a fourth of a day?	Work Space
There are six hours in a fourth of a day.	24 ÷ 4 = 6

Day 25

| Gabrielle has 10 pink ribbons. She has 8 fewer green ribbons than pink ribbons. She has 16 more purple ribbons than green ribbons. How many ribbons does Gabrielle have in all?

Gabrielle has 30 ribbons in all. | Work Space

Green: 10 - 8 = 2
Purple: 2 + 16 = 18
All: 10 + 2 + 18 = 30 |

Day 26

Ethan saved $54.00. He earned $35.00 for cutting the lawn. Then he spent $10.00 on a pair of goggles, $15.00 on an iPad cover, $14.75 for a sports cap, and $25.00 for a blue scarf. How much money does Ethan have left? Ethan has $24.25 left.	Work Space He saved 54 + 35 = 89 He spent 10 + 15 + 14.75 + 25 = 64.75 $89.00 - 64.75 = $24.25

Day 27

In Michael's desk drawer, there are 14 yellow markers. There are 4 more black markers than yellow markers, and there are 11 more green markers than black markers. How many markers are in Michael's desk drawer altogether? There are 61 markers in Michael's desk drawer.	Work Space 4 + 14 = 18 18 + 11 = 29 14 + 18 + 29 = 61

Day 28

Jason planted four types of fruit trees in his vineyard. He planted 37 of each type of fruit. How many fruit trees did Jason plant in his vineyard? Jason planted 148 fruit trees in his vineyard.	Work Space 37 X 4 = 148

Day 29

The Simmons' Explorer has 816 campers. If each cabin can hold 8 campers, how many cabins will they need? They will need 102 cabins.	Work Space $816 \div 8 = 102$

Day 30

Elizabeth was collecting sparkling marbles. She got 23 marbles in a fish bowl, and Emmanuel gave her 8 marbles. Her dad gave her 14 marbles. Elizabeth gave 17 marbles to her cousin Alex. How many marbles did Elizabeth have left? Elizabeth had 28 marbles left.	Work Space $23 + 8 + 14 = 45$ $45 - 17 = 28$

Day 31

George wants to triple the size of his work area. He has an area of 123 feet. How much work area does George want? *Do you understand the question?* *Make a plan to solve the problem.* *Carry out your plan to solve the problem.* *Check your answer.*	Work Space The area that George wants. Write an equation. Make a diagram. $123 \times 3 = 369$ feet $123 + 123 +$ $123 - 369$ Look back and review.

Day 32

Rosie has a rectangular flower garden that is 4 ft long and 6 ft wide. What length of fencing does she need to fence in her flower garden? Rosie needs 20 ft. of fence for fencing in her flower garden. ***Do you understand the question?*** ***Make a plan to solve the problem.*** ***Carry out your plan to solve the problem.*** ***Check your answer.***	Work Space 2 X (4ft + 6 ft) = 20 ft

Day 33

Dr. Now bought a one-year fitness contract for $ 420.00. If she makes monthly payment for a year how much is each payment: Dr. Now's monthly payment will be $ 35.00.	Work Space $420 ÷ 12 = $35

Day 34

Your mother wants new carpet in her bedroom. Her bedroom is 12m by 12m rectangle. How much carpet does she needs to buy to cover her bedroom floor? She needs to buy 144 sq m carpet to cover the entire bedroom floor.	Work Space 12 X 12 = 144

Day 35

A daycare center teacher ordered 69 boxes of popsicle sticks. There were 85 popsicle sticks in each box. How many popsicle sticks did the teacher order in all? The teacher ordered 5,865 popsicle sticks in all.	Work Space 69 X 85 = 5,865

Day 36

The cooks at the shelter are in need of 654 potatoes. There are 6 potatoes in each sack. How many sacks of potatoes do they need? They need 109 sacks of potatoes.	Work Space 654 ÷ 6 = 109

Day 37

The school principal has $6,918 to spend on replacing tiles for the classroom. If the cost of each tile is $3.00, how many tiles can the principal buy? The principal can buy 2,306 tiles.	Work Space 6,918.00 ÷ 3 = 2,306

Day 38

A district of teachers needs to go to a workshop in a nearby city. There are 780 teachers in the district. If each passenger van carries 15 teachers, how many passenger vans does the district need? The district needs 52 passenger vans.	Work Space 780 ÷ 15 = 52

Day 39

A cheerleader squad has $829.00 to spend on buttons for their fans. Each button costs $0.85. Approximately how many buttons can the squad purchase? The squad can purchase approximately 975 buttons for their fans.	Work Space $829 ÷ $0.85 = 975.29

Day 40

At the Carolina Fair, a group of 554 seniors wants to ride the giant tagmahall ride. If each booth on the Tagmahall can hold 9 seniors in each car, approximately how many cars will the seniors need? The seniors will need approximately 62 cars.	Work Space 554 ÷ 9 = 61.56

Day 41

Your dad spent $287.00 on a diamond necklace and $878.00 on a set of diamond earrings. How much money did he spend in all for your mother's birthday jewelry? He spent $1,165 for the birthday jewelry.	Work Space $287 + $878 = $1,165

Day 42

Before lunch, Jason has a science class for 1 hour and 40 minutes. Lunch lasts 45 minutes and ends at 1:00 p.m. What time does his science class start? Jason's science class starts at 10:35 a.m.	Work Space Jason's science class starts at 10:35 a.m.

Day 43

You need to drink 7 ¼ cups of water and 3/4 cups of milk every day. Approximately how much liquid do you have to drink daily? I have to drink around 8 cups of liquid daily.	Work Space 7 ¼ + ¾ = 7 + 4/4 = 7 + 1 = 8

Day 44

The class has 24 students, and ¾ of them are boys. How many are boys in the class? There are 18 boys in the class.	Work Space ¾ of 24; 24 ÷ 4 = 6; 6X3=18; ¾ of 24 = 18

Day 45

The cook made 48 cupcakes. One-eighth of the cupcakes have green icing. How many cupcakes are green? There are 6 green cupcakes.	Work Space 1/8 of 48 48 ÷ 8 = 6

Day 46

Three-sixths of the 18 children on the playground are playing kickball. How many children are playing kickball? There are 9 children playing kickball.	Work Space 3/6 of 18 18 ÷ 6 = 3 3 X 3 = 9 3/6 of 18 = 9

Day 47

There are 36 students in the cafeteria eating lunch. Three-fourths of the students are eating cafeteria lunch. How many students are eating the cafeteria lunch? There are 27 students eating the cafeteria lunch.	Work Space ¾ of 36 36 ÷ 4 = 9 9 X 3 = 27 ¾ of 36 = 27

Day 48

There are 50 people waiting for the greyhound bus. One-fifth of them have two or more pieces of luggage. How many people waiting have more than two pieces of luggage? There are 10 people with two or more pieces of luggage.	Work Space 50 ÷ 5 = 10

Day 49

One-fifth of the 895 students at school wear glasses. How many students wear glasses? There are 179 students who wear glasses.	Work Space 895 ÷ 5 = 179

Day 50

There are 750 houses in our community. Two-fifths of the houses are made of bricks. How many houses are made of bricks? There are 300 houses that are made of bricks.	Work Space 750 ÷ 5 = 150 150 X 2 = 300

Day 51

Renting a bike at the amusement park costs $12.00 per hour. A family of four rented 4 bikes. How much will they pay to rent the bikes for 6 hours? They will pay $192 to rent the bikes for 6 hours.	Work Space $12.00 X 6 = 48.00 $48.00 X 4 = $192.00 or 48 + 48 + 48+ 48 = $192

Day 52

Jason earned $ 25.00 for grooming the dog. He earned $35.00 for cutting the lawn. His brother Michael earned $ 65.00 for replacing a ceiling fan. How much money did Jason earn in all? He earned $60 in all.	Work Space $25.00 + $35.00 = $60.00

Day 53

Janee' walks for 30 minutes each day. She does aerobics for 25 minutes each day. How many minutes does Janee' walk and do aerobics in 7 days? Janee walks and does aerobics for 385 minutes in 7 days.	Work Space 30 + 25 = 55 55 minutes X 7 days = 385 minutes or 6 hours and 25 minutes

Day 54

Tickets to the amusement park cost $12.00 for children and $15.00 for adults. How much will it cost for 6 adults and 9 children to attend the amusement park? *Talk with partner the correct order of the operations you need to use to solve the problem: multiply and then subtract; subtract and then multiply; multiply and then add; or add and then multiply*	Work Space $12 X 9 = $108 $15 X 6 = $90 $108 + $90 = $198

Day 55

It takes Michael 53 minutes to walk from his house to the store. If Michael needs to be at the store for work at 8:30 a.m., what time does he need to leave his house? Michael needs to leave his house by 7:37 a.m. to get to work on time.	Work Space 7:37 a.m.

Day 56

Mrs. Simmons bought a purse and a pair of shoes at the Department Store. She paid $59.00 for the shoes and $29.99 for the purse. The clerk gave her $1.01 change. How much money did she give the clerk? She gave the clerk $90.00.	Work Space $59.00 + $29.99 = $88.99 $88.99 + $1.01 = $90.00

Day 57

Reduce the fraction 4/12 to its lowest term.	Work Space 4/12 ÷ (4/4) = 1/3

Day 58

Reduce the fraction 6/8 to its lowest term.	Work Space 6/8 ÷ 2/2 = ¾

Day 59

What is the area of the rectangle? 10cm 6cm	Work Space A = l x w 10 cm x 6 cm = 60 cm^2

Day 60

What is the perimeter of the square?	Work Space
☐ 8 m	P = 4s P = 4 x 8 P = 32 m

Day 61

What is the perimeter of the rectangle?	Work Space
6cm ☐ 7cm	P = 2l + 2w P = (2x6) + (2x7) P = 12 + 14 P = 26 cm

Day 62

What is the area of the figure below?	Work Space
☐ 8 in. 3 in.	A = l x w A = 8 x 3 A = 24 in

Day 63

Draw a right angle.	Work Space
	∟

Day 64

Draw an obtuse triangle.	Work Space
	(obtuse angle drawing)

Day 65

A ticket to the talent show costs $3. How much will it cost if 5 people go? 7 people? 9 people?	Work Space
	$3 X 5 = $15
	$3 X 7 = $21
	$3 X 9 = $27

Tickets	1	2	3	4	5	6
Cost	$3	$6	$9	$12	$15	$18

Day 66

Reduce 8/12 to the lowest term.	Work Space
	8/12 ÷ 4/4 = 2/3

Day 67

What two fractions are equivalent to 4/5?	Work Space
	4/5 = 8/10
	= 16/20

Day 68

Divide this pizza into 8 equal parts. 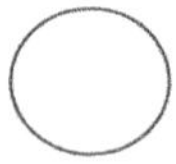 ◯ Shade in 5/8 of the pizza.	Work Space

Day 69

The cat's birthday cake has ¼ cup of milk and ¾ cup of water. How much liquid is needed in the cake? 1 cup of liquid is needed in the cake.	Work Space ¼ + ¾ = 4/4 = 1 cup of liquid

Day 70

James spent 3/6 of an hour studying for his social studies exam and 5/6 of an hour studying for his French test. How much longer did James spend studying for French than for social studies? He spent 20 minutes more studying French than studying social studies.	Work Space 3/6 of 60 minutes = 30 minutes 5/6 of 60 minutes = 50 minutes

Day 71

Thea bought ¾ lbs. of green bananas and 2/4 lbs. of yellow (ripe) bananas. How many pounds of bananas did Thea buy in all? Thea bought 1 ¼ lbs. of bananas in all.	Work Space ¾ + 2/4 = 5/4 = 1 ¼

Day 72

Helen baked 252 cookies. She wants to give all six of her grandchildren the same number of cookies. How many cookies should each grandchild get? Helen should give each grandchild 42 cookies.	Work Space 252 ÷ 6 = 42

Day 73

Jason received $ 85.00 from his part-time job. He went to a football game, a movie, and the game room. He had $12.50 left. How much did he spend on entertainment? Jason spent $72.50 on entertainment.	Work Space $ 85.00 - $ 12.50 $ 72.50

Day 74

Last week, Ethan made $79.00 doing chores. He decided to wash cars at the pit stop. He now has $130.00. How much money did he make washing cars? Jason made $51.00 washing cars.	Work Space $130.00 - $79.00 ――――― $ 56.00

Day 75

Frederick has 85 watermelons left at his vegetable stand. He sent his workers to get more watermelons to stock for the afternoon. Now 105 watermelons are at his stand. How many did the workers bring from the field? Frederick's workers brought 20 watermelons from the field.	Work Space 105 - 85 20

Day 76

Draw a clock with the time 1:40 c. What time will it be in 2 hours and 30 minutes? 4:10 d. What time will it be in 3 hours and 15 minutes? 4:55	Work Space a. 4:10 b. 4:55

Day 78

Are the lines below parallel? Explain	Work Space No. Parallel lines stay the same distance apart. They never intersect.

Day 79

Are the lines below parallel? Explain.	Work Space
∥	Yes. Parallel lines stay the same distance apart.

Day 80

Are the lines below intersecting? Explain.	Work Space
✕	Yes. The lines meet at one point.

Day 81

This shape has how many vertices? Explain.	Work Space
⯃	Vertices are the same as corners. The shape has 8 vertices.

Day 82

How many sides does this shape have? ⬡ This shape has 6 sides.	Work Space — Sides are the straight lines that make up a shape. This shape has <u>6</u> sides.

Day 83

What is the unknown number?	Work Space
7 + r = 11	7 + r = 11
	7 + r -7 = 11 -7
r = 4	r = 4

Day 84

Johnny has $24.00 to spend on six ink pens. After buying the pens, he had $12.00 left. How much did each ink pen cost?	Work Space
	$24 - 12 = 12
	12 ÷ 6 = $2
Each ink pen cost $2.	

Day 85

There were 324 students going to the zoo. The students filled six buses. How many students were on each bus?	Work Space
	324 ÷ 6 = 54
	54
There were 54 students on each bus.	X 6
	324

Day 86

After having lunch at a restaurant, Dee, Rosemary, and Robin decided to evenly split the bill. If each person paid $ 19.00, what was the total bill for the lunch?	Work Space
	$19
	X 3
	$57
The total bill for the lunch was $57.	

Day 87

Converting: How many minutes and hours are in 205 minutes? *Remember: 60 minutes = 1 hour* There are 3 hours and 25 minutes in 205 minutes.	Work Space 3 hours and 25 minutes

Day 88

Converting: How many minutes and hours are in 159 minutes? *Remember: 60 minutes = 1 hour* There are 2 hours and 39 minutes in 159 minutes.	Work Space 2 hours and 39 minutes

Day 89

Converting: How many minutes are in 5 hours? *Remember: 60 minutes = 1 hour* There are 300 minutes in 5 hours.	Work Space 60 <u>X 5</u> 300 minutes

Day 90

Solve the equation: 3 = 6 t - 12 – 3 t = 3	Work Space 3= 6 t - 12 - 3 3 = 6 t - 15 + 15 3 + 15 = 6 t 18 = 6t 18/6 = 6t / 6 3 = t

Day 91

Solve the equation: 36 = - p -5 p p = 6	Work Space 36 = -p -5p 36 = - 6 p 36 / 6 = -6 p/-6 6 = p

Day 92

Solve equation: q + 8 - 2 = 9	Work Space q + 8 - 2 = 9 q + 6 = 9 q + 6 - 6 = 9 - 6 q = 3

Day 93

Make a bar graph with 7 classmates' favorite vegetable	Work Space
Favorite Vegetables graph with columns: spinach, green beans, cabbage, carrots, cucumber	The answers will vary.
What is the most favorite vegetable? _____ What is the least favorite vegetable? _____	

Day 94

Create a graph of 10 classmates' favorite subjects:	Work Space
Favorite Subject graph	The answers will vary.

Day 95

Make a bar graph using a cup of colorful marshmallows and create two questions about your graph.	Work Space
Graphing grid	

Day 96

Solve the equation: 4 X 27 ÷ 3 = 36	Work Space 4 x 27 ÷ 3 4 X 9 36

Day 97

Solve the equation: (20 X 3) ÷ 3 = 20	Work Space (20 X 3) ÷ 3 (60) ÷3 60 ÷ 3 = 20

Day 98

Solve the equation: 3.07 + 5. 4 - 2.1 = 6.37	Work Space 3.07 + 5. 4 - 2.1 = 6.37

Day 99

Using decimals: The vet is 7.8 kilometers west of the gas station and 25.6 kilometers east of the beauty salon. How far apart are the gas station and the beauty salon? The gas station and the beauty salon are 17.8 kilometers apart.	Work Space 25.6 – 7.8 = 17.8

Day 100

Using decimals: The mechanic shop is 71.6 miles west of the donut shop and 54.5 miles east of the barber shop. How far is the donut shop from the barber? The donut shop is 17.1 miles from the barber shop.	Work Space 71.6 − 54.5 = 17.1

Day 101

Rachael is reading a mystery novel that has 581 pages. She wants to finish reading the novel in seven days. How many pages does she need to read each day? She needs to read 83 pages each day.	Work Space 581 ÷ 7 = 83

Day 102

The Parent and Teacher Organization (PTO) bought 14 boxes of lead pencils. Each box has 18 lead pencils. The PTO paid $6.00 per box. If the PTO sells each lead pencil for $0.50, how much profit will the PTO make? The PTO will make $ 42.00 profit.	Work Space 14 X $6.00 = $84.00 18 X $ 0.50 = $9.00 14 X $9.00 = $ 126.00 $ 126.00 - 84.00 = $42.00

Day 103

Jason and his friends play baseball on Saturday. The game begins at 2:30 p.m. It will take Jason and his friends 40 minutes to get to the field. Approximately, what time do they need to leave in order to get to the baseball field 30 minutes before the game begins? They need to leave at 1:20 p.m.	Work Space 1:20 p.m.

Day 104

Michael's football practice began at 11:00 a.m. The team practiced offense for 65 minutes, and 1 hour and 45 minutes on defense techniques. What time did Michael's practice end? Michael's practice ended at 1:50 p.m.	Work Space 1:50 p.m.

Day 105

Elizabeth went to the park at 2:35 p.m. She played in the sandbox for 30 minutes, the slides for 10 minutes, and the swings for 15 minutes. What time did Elizabeth leave the park? Elizabeth left the park at 3:30 p.m.	Work Space 3:30 p.m.

Day 106

On Tuesday, Emmanuel had a student council meeting for 45 minutes after class, followed by keyboarding music lessons for 1 hour and 10 minutes. If Emmanuel's classes end at 3:15 p.m., what time was he finished with his music lesson? He finished his music lesson at 5:10 p.m.	Work Space 5:10 p.m.

Day 107

John bought 43 cases of carnation flowers. There were 35 carnations in each case. How many carnations did John buy? John bought 1,505 carnations.	Work Space 43 x 35 = 1,505

Day 108

The school district has 65 buses. There are 39 seats on each bus. Approximately how many children can the buses transport to school? The buses can transport 2,535 children to school.	Work Space 65 x 39 = 2,535

Day 109

	Work Space

At Elizabeth's birthday party, the girls ate 2 ½ pizzas and the boys ate 4 ¾ pizzas. How many pizzas did the boys and girls eat in all?
 	
The boys and the girls ate 7 ¼ pizzas in all.
	1 ½ = 2/4
	     4 ¾ = ¾
	―――――
	    6 5/4
	         7 ¼

Day 110

	Work Space
Mr. Johnny bought 5 gallons of paint. He used only 1 ¾ gallon. How much paint does Mr. Johnny have left? Mr. Johnny has 3 ¼ gallons of paint left.	5 - 1 ¾ = 3 ¼

Day 111

	Work Space
My German Shepherd is 6 ½ years old, and my Cocker Spaniel is 2 ½ years younger than the German Shepherd. How old is my Cocker Spaniel? The Cocker Spaniel is 4 years old.	6 ½ - 2 ½ = 4

Day 112

Shade the parts to illustrate the decimal number 0.4.	Work Space

Day 113

Shade the parts to illustrate the decimal number 0.74.	Work Space

Day 114

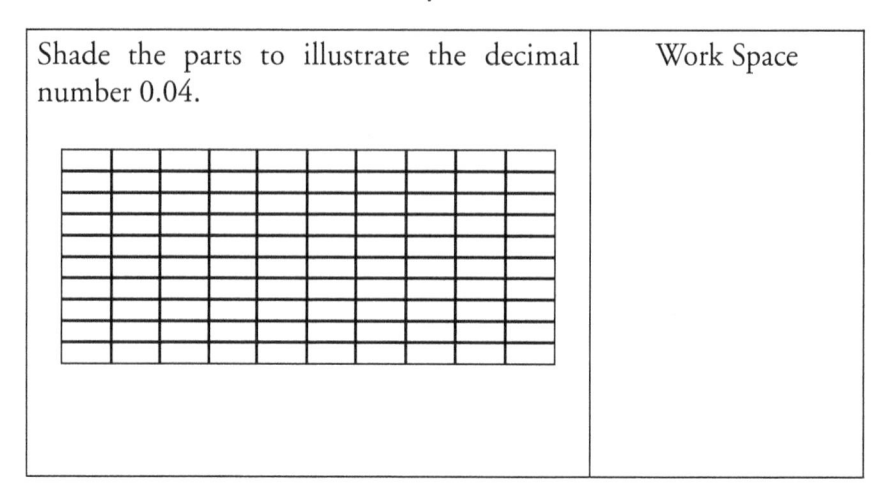

Shade the parts to illustrate the decimal number 0.04.	Work Space

Day 115

Decimals: Janee bought 5 .1 meters of cloth to cover a scrapbook. She used 4.3 meters of the cloth. How much cloth did Janee have left? Janee had 0.8 meters of cloth left.	Work Space $5.1 - 4.3 = 0.8$

Day 116

Jason had 0.9 grams of salt. He used only 0.5 grams of salt to make chicken fritters. How much salt did Jason have left? Jason had 0.4 grams of salt left.	Work Space $0.9 - 0.5 = 0.4$

Day 117

	Work Space
Probability: Use words like "certain," "unlikely," "impossible," and "probably" to solve problems. Illustrate 3 blue marbles and 2 yellow marbles. How likely is it that you will pick a blue marble out of a bag? Explain.	Probably

Day 118

	Work Space
Probability: Use words like "certain," "unlikely," "impossible," and "probably" to solve problems. Illustrate 5 yellow rocks and 4 brown rocks. How likely will you pick a yellow rock?	Probably

Day 119

	Work Space
Place Value: Looking at the number 3,485, which digit is in the tens place? 8 is in the tens place	8

Day 120

	Work Space
Even Numbers: Write the even number from 44 to ___. 44, ___, ___, ___, ___, ___, ___, ___, ___, ___, ___, ___, ___, ___, ___, ___, ___, ___, ___, ___	46, 48, 50, 52, 54, 56, 58, 60, 62, 64, 66, 68, 70, 72, 74, 76, 78, 80, 82

Day 121

Odd Numbers: Write the odd numbers from 1 to ___. 1, ___	Work Space 3, 5, 7, 9, 11, 13, 15, 17, 19, 21, 23, 25, 27, 29, 31, 33, 35, 37, 39, 41, 43, 45, 47, 49, 51, 53, 55, 57

Day 122

Counting by 2s. 50, ___	Work Space 52, 54, 56, 58, 60, 62, 64, 66, 68, 70, 72, 74, 76, 78, 80, 82, 84, 86, 88, 90, 92, 94, 96, 98, 100, 102, 104

Day 123

Counting by 4s from 44. 44, ___	Work Space 48, 52, 56, 60, 64 68, 72, 76, 80, 84 88, 92, 96, 100, 104, 108, 112, 116, 120, 124, 128, 132, 136, 140, 144 148, 152

Day 124

Counting by 10s from 43.	Work Space
43, ___	53, 63, 73, 83, 93, 103, 113, 123, 133, 143, 153, 163, 173, 183, 193, 203, 213, 223, 233, 243, 253, 263, 273, 283, 293, 303

Day 125

Counting by 10s from 112.	Work Space
112, ___	122, 132, 142, 152, 162, 172, 182, 192, 202, 212, 222, 232, 242, 252, 262, 272, 282, 292, 302, 312, 322, 332, 342, 352, 362, 372, 382

Day 126

Counting by 5s from 102.	Work Space
102, ___	107, 112, 117, 122, 127, 132, 137, 142, 147, 152, 157, 162, 167, 172, 177, 182, 187, 192, 197, 202, 207, 212, 217, 222, 227, 232, 237

Day 127

Number Words: Write the number word for 587. Five hundred eighty-seven	Work Space five hundred eighty-seven

Day 128

Number Words: Write the number word for 1,802. One thousand eight hundred two	Work Space one thousand eight hundred two

Day 129

Ethan bought a 5-lb. bag of apples for $4.38, and he gave the cashier $5.00. How much change will he get? He will get a change of $0.62.	Work Space $5.00 − 4.38 = 0.62

Day 130

You went to the store to get a gallon of vanilla ice cream. The ice cream cost $5.89. Your mom gave you 3 one-dollar bills, 8 quarters, 5 dimes, and 6 pennies. How much money did your mom give you? $5.56 Do you have enough money to get the ice cream? No How much money do you have left or how much more do you need? I need $0.33 more.	Work Space 3 one-dollar bills = $3.00 8 quarters = $2.00 5 dimes = $0.50 6 pennies = $0.06

Day 131

We are planting 48 sweet potato plants and would like them to be in 6 rows. How many sweet potato plants need to be in each row?	Work Space 48 ÷ 6 = 8

Day 132

At your lemonade stand, you sold lemonade for $0.15 a cup. You made $18.00. How many cups of lemonade did you sell? I sold 120 cups of lemonade.	Work Space $18.00 ÷ 0.15 = 120

Day 133

The PTO sold T-shirts for $10.00 each to 856 students at school. Everyone purchased a T-shirt for $10.00. How much money did the PTO make selling the T-shirts? $8,560 The PTO paid $6.25 for each T-shirt. How much profit did the PTO make? $3,210	Work Space 856 X 10 $8,560 (made) 6.25 X 856 $5,350 (paid) $8,560 - 5,350 $3,210 (profit)

Day 134

	Work Space
Mrs. Frederick made a batch of cookies. There were 24 cookies in each batch. She ate ¼ of the batch. Then Mrs. Frederick baked three more batches. How many cookies does Mrs. Frederick have now? Mrs. Frederick has 90 cookies now.	24 ÷ ¼ = 6 24 X 4 96 96 – 6 =90

Day 135

	Work Space
Emmanuel's grades in science were 89, 93, 87, 77, 100, and 85. What is his average grade in science? Emmanuel's average grade in science is 88.5.	88.5

Day 136

	Work Space
Angie wants to give the server a 15% tip. Her lunch receipt is $18.89. How much will Angie give the server? Angie will give the server $ 2.83.	18.89 X .15 9445 +18890 2.8335 $ 2.83

Day 137

A rectangular court measures 12 ft. X 5 ft. What is the area of the court? The area of the court is 60 sq. ft.	Work Space A = l X w 12 X 5 60 sq. ft

Day 138

A square room measure 8 ft. on the right side. What is the perimeter? The perimeter of the room is 36 sq. ft.	Work Space A square has four equal sides. P = 6 + 6 + 6 + 6 = 36 sq. ft.

Day 139

George is putting new carpet in his bedroom. His bedroom is 21 ft. X 17 ft. What is the area of his bedroom? The area of his bedroom is 357 sq. ft.	Work Space A = l X w 21 X 17 147 21 357 sq. ft.

Day 140

Dr. Dargan has a rose garden that is 8 m by 12 m. One bag of manure can cover 12m. How many bags of horse manure will Dr. Dargan use? Dr. Dargan will use 8 bags of horse manure.	Work Space 12 X 8 — 96 m 96 ÷ 12 = 8 bags

Day 141

Elapsed time: If it is 8:15 a.m., what time will it be in 7 hours and 15 minutes? It will be 3:30 p.m. in 7 hours and 15 minutes.	Work Space 3:30 p.m.

Day 142

Elapsed time: If it is 1:45 p.m., what time will it be in 7 hours and 30 minutes? It will be 9:15 p.m. in 7 hours and 30 minutes	Work Space 9:15 p.m.

Day 143

Mark started baking some chocolate chip cookies at 4:35 p.m. It took 35 minutes to prepare the ingredients and 2 hours and 10 minutes to bake all the cookies. When did Mark finish baking? Mark finished baking at 7:20 p.m.	Work Space 7:20 p.m.

Day 144

Ethan and Elizabeth were watching a video at 5:35 p.m. The video lasted for 1 hour and 45 minutes. Then Ethan and Elizabeth played basketball for 35 minutes. What time did Elizabeth and Ethan finish playing basketball? Elizabeth and Ethan finished playing basketball at 7:55 p.m.	Work Space 7:55 p.m.

Day 145

What shape has three equal sides? Illustrate and name the shape.	Work Space Equilateral triangle △

Day 146

What shape has four equal sides and four corners? Illustrate and name the shape.	Work Space Square □

Day 147

What shape has six sides and six angles? Illustrate and name the shape.	Work Space Hexagon ⬡

Day 148

What shape has eight sides? Illustrate and name the shape.	Work Space octagon ⬡

Day 149

Convert the number of miles it takes to equal 15,840 feet. Three miles equals 15,840 feet.	Work Space 3 1 mile = 5,280 feet

Day 150

How much time has elapsed between the first and second time? *First* *Second* *Elapsed Time* 1:45 2:30 45 minutes 4:15 5:46 1 hour 31 minutes 9:35 9:45 10 minutes	Work Space 45 minutes 1 hour 31 minutes 10 minutes

Day 151

Thea has 895 erasers in a box. She sold 186 erasers. How many does she have left? She has 709 erasers left.	Work Space 895 − 186 = 709

Day 152

Lillian has 834 dimes in her piggy bank. She spent 479 of her dimes. How many dimes does she have now? She has 355 dimes now.	Work Space 834 – 479 = 355

Day 153

Lula sold 819 cooking magazines on Monday, and sold an additional 749 magazines on Tuesday. How many more magazines did she sell on Monday than she sold on Tuesday? She sold 70 more magazines on Monday than she did on Tuesday.	Work Space 819 – 749 = 70

Day 154

A toy factory made 589,877 trucks and 539,227 dolls. How many more trucks than dolls did the toy factory make? The factory made 50,650 more trucks than dolls.	Work Space 589,877 – 539,227 = 50,650

Day 155

During the midnight shift, the workers made 5,277 small bobby pins and 3,456 large bobby pins. How many bobby pins did the workers make during the midnight shift? The workers made 8,733 during the midnight shift.	Work Space 5,277 + 3,456 = 8,733

Day 156

	Work Space
Last year, a candy factory made 6,899 chocolate Easter eggs. This year, the candy factory made 10,100 chocolate Easter eggs. How many more chocolate eggs did they make this year than last year? They made 3,201 more chocolate eggs this year than last year.	10,100 - 6,899 = 3,201

Day 157

	Work Space
Simmons and Brown Enterprise sold 6,990 paperback books. The company also sold 3,011 eBooks. How many books did Simmons and Brown Enterprise sell in all? Simmons and Brown enterprise sold 10,001 books in all.	6,990 +3,011 10,001

Day 158

	Work Space
The Crum's Foundation cut down 4,660 trees on the north side of the estate. They cut down 5,896 more from the south side of the estate. How many more trees were cut down from the south side than the north side? There were 1,236 more trees cut down from the south side than the north side.	5,896 – 4,660 = 1,236

Day 159

Wilhemenia wants to build a fence around her backyard. Wilhemenia's backyard is 15 meters long and 15 meters wide. It costs $7.00 per meter to install the fence. How much will it cost to put fence around Wilhemenia's backyard? It will cost $1,575.00 to put a fence around Wilhemenia's backyard.	Work Space 15 m X 15 m = 225 sq m 225 sq m X $ 7.00 = $ 1,575.00

Day 160

The CEO wants to install a new floor in the office at the RCRC. The office measures 9 meters wide and 12 meters long. The flooring costs $8.00 per square meter. How much will the CEO pay for the new floor in the office? The CEO will pay $864 for the new floor in the office.	Work Space 12 X 9 = 108 108 X $8.00 = $864

Day 161

Patterns: 1, 3, 9, ___, 81, 243 1, 2, 4, ___, 16, 32 1, 3, 9, ___, 81	Work Space 27, 8 27

Day 162

The Ridgeville Community Resource Center used a grant to buy 13,456 books for preschool children. The center distributed 8,045 of the books. How many books does the center have left? The center has 5,411 books left.	Work Space 13,456 − 8,045 = 5,411

Day 163

The staff at the Native American Center treated 879 clients this month. Last month the staff treated 1,245 clients. How many more clients did the staff treat last month compared to this month? The staff treated 366 more clients last month than this month.	Work Space 1245 − 879 = 366

Day 164

What is the median of each of the two sets of numbers? Set 1: 3 5 5 7 8 9 9 Set 2: 1 2 2 4 5 6 8 The median of set 1 is 7. The median of set 2 is 4.	Work Space 7 4 When numbers are arranged from the least to the greatest the median is the number in the middle.

Day 165

What is the mean of each of the two sets of numbers? Set 1: 1 6 7 3 8 9 3 3 Set 2: 3 5 4 9 7 9 6 5 The mean of set 1 is 5. The mean of set 2 is 6.	Work Space 5 6 The mean is the average of the numbers.

Day 166

What is the median of each of the two sets of numbers? Set 1: 8 8 7 6 5 4 1 Set 2: 9 6 4 7 8 3 2 The median of set 1 is 6. The median of set 2 is 6.	Work Space 6 1 4 5 6 7 8 8 6 2 3 4 6 7 8 9 The median is the number in the middle; therefore, the numbers must be arranged from the least to the greatest. Then find the number in the middle.

Day 167

What is the mode of each of the two sets of numbers? Set 1: 9 5 6 5 3 5 6 4 Set 2: 0 0 7 9 2 1 0 3 The mode of set 1 is 5. The mode of set 2 is 0.	Work Space 5 0 Mode is the number that appears the most.

Day 168

What is the mode of each of the two sets of numbers? Set 1: 8 9 6 8 5 8 Set 2: 7 7 7 9 9 3 3 The mode of set 1 is 8. The mode of set 2 is 7.	Work Space 8 7 Mode is the number that appears the most.

Day 169

What is the range of each of the two sets of numbers? Set 1: 5 9 4 8 9 4 3 Set 2: 3 7 5 10 4 2 7 9 The range of set 1 is 6. The range of set 2 is 8.	Work Space 9 - 3 = 6 10 - 2 = 8 The range is the difference between the largest to the smallest number.

Day 170

Angie bought a bag of speckled beans. The clerk asked for $1.79 for the bag of speckled beans. What coins can she give the clerk to pay the cost of the speckled beans?	Work Space The answers may vary.

Day 171

Wanda bought a container of eggs for $2.97. She gave the cashier $5.00. How much change will she get back? She will get a change of $2.03.	Work Space $5.00 − 2.97 = $2.03

Day 172

Jason bought a pack of fat free cheese for $3.99. He has 2 dollars, 4 quarters, and 5 dimes. How much change will Jason get back or how much more money will he need to buy the fat-free cheese? He will need $0.49 more to buy the fat-free cheese.	Work Space $2.00 + $1.00 + $0.50 = $3.50

Day 173

Write a word problem with your partner and design a strategy to get an answer.	Work Space

Day 174

Write a word problem with your partner and design a strategy to get an answer.	Work Space

Day 175

Write a word problem with your partner and design a strategy to get an answer.	Work Space

Day 176

Write a word problem with your partner and design a strategy to get an answer.	Work Space

Day 177

Write a word problem with your partner and design a strategy to get an answer.	Work Space

Day 178

| Write a word problem with your partner and design a strategy to get an answer. | Work Space |

Day 179

| Write a word problem with your partner and design a strategy to get an answer. | Work Space |

Day 180

| Write a word problem with your partner and design a strategy to get an answer. | Work Space |

CPSIA information can be obtained
at www.ICGtesting.com
Printed in the USA
LVHW042149270119
605462LV00001B/122/P